Though the words
in this treatise
might modify your
view of yourself
and the world,

they do not constitute
a philosophy per se
(though some may say
it opens the door to one);
nor do they in any way
constitute a religion,
or an ideology, or a
prediction, or prophecy;
nor do they allude to
fate or destiny;

they are, basically,
simple, objective
statements of fact
that you will
acknowledge, or
deny, according
to how the *you*
that you are
came to be.

The Tiny Truth
That Governs
Everything
(EXPANDED EDITION)

ISBN 978-1543168839

fabianmelmelgar@gmail.com

Also by
Fabian (Mel) MelGar:

**The One Truth
That Governs
Man, The Gods,
The Heavens
And The Universe**
© 2006
(Early version of
The Tiny truth
That Governs
Everything)

**Paint The Beast
Pretty** (a cerebral
mystery novel)
© 2013

FabianMelMelGar

EXPANDED EDITION

THE
TINY
TRUTH

THAT
GOVERNS

EVERY THING

The *ifs*, *ands*, and *buts* about it:

If the tiny truth of this treatise were not true, you could sing no song, you could smell no rose; you could marvel not at the rise of the sun or the sweep of the moon; if it were not true, no wheel could turn, no gear could grind, no heart could quiver, no mind could wonder; if it were not true, there would be no me and there could be no thee; if it were not true, there could be no clouds and there could be no rain, there could be no streams and there could be no seas; and if it were not true there could be no butterflies afluttering or elephants atrumpeting; nor could there be a little blue orb warming up close to a large ball of roiling hot embers in what is an immense, mostly empty and cold universe; but this tiny truth, which makes that little blue planet and our place on it possible, is hardly known, though knowledge of it might change us and our world, it might make us more accepting each of the other, it might make us more open, more caring, more rational, less dogmatic, less bigoted; it could nudge our world toward a nobler destiny, a more equalitarian future where everyone mattered and had the opportunity to achieve to the best of their abilities, a future where everyone would contribute to the creation of a world that could be the envy of the universe, if they but knew us.

So, here, whether you suspect that

you know everything there is to know

about it, or you are—or were till now—

totally unaware of its existence;

and with the object of making it

more known and understood

to all who prize truth,

the simple ins and outs of the

tiny truth that governs everything:

To begin:

There are many theories about man,

and many beliefs about the gods;

many tenets about the heavens,

and many facts about the universe;

but there is only one truth

that governs man *and* the gods, *and*

the heavens, *and* the universe;

and it stands on two simple principles:

one is that *all* moments in time,

whether eventful or not, whether

regular or random, whether

understood or not, and whether

predictable or not, are caused;

the other, a bit less simple, is that

all motion proceeds in the direction

where the cause precedes the effect,

regardless of the direction of time;

and these two principles

are embodied in this one tiny truth:

**Everything as it is
at this moment**

was caused by
everything as it was
a moment ago.

That may seem an innocuous little truth;

but consider that everything

as it was at *that* moment was caused by

everything as it was the moment before *that*,

and everything as it was at *that* moment

was caused by everything as it was the

moment before *that,* and the moment

before *that*, and the moment before *that;*

and the moment before *that;*

which means:

that everything as it is at this moment,

was caused by everything as it was at any,

and all, of the unknowable, uncountable,

trillions, upon trillions, upon trillions, upon

trillions of moments that have already been

—the fact, for example, that you are reading

this at this moment was causally determined

by everything as it was twenty six trillion,

seventy five billion, six million, two hundred

twenty eight thousand, one hundred sixty one

moments ago; and also by everything as it was

just thirty three moments ago, and one

moment ago, and all the moments between.

It cannot be otherwise.

And
everything
means

EVERY

R

THING

—past,
present
and
future—
including, but not limited to . . .

the 7,400,000,000+
humans on Earth,

THE 2,000,000,000,000,000,000,000+++ SUNS IN OUR GALAXY,

every
bounce
of a ball,

every
thought
in your
head.

Nothing is excluded,
no moment of time,
no random event,
no man, no god,
no grain of sand,
no living thing,
no anything;

11

And just as *everything* means *everything*,
caused means *caused*, and
determined means *determined*;

and the tiny truth that *everything as it is at this
moment was caused by everything as it was a moment ago,* means, exactly that: that *everything as it
is at this moment was causally determined by
everything as it was a moment ago,* and if we
were to look *back* at that ago moment we could
see the unequivocal truth of that.

If, however, we were to look *forward* in time:
while we could say —unequivocally— that **everything as it is at this moment will causally
determine** everything as it will be a moment from
now; we could not say —unequivocally— because
of the complexity and randomness of events,
*that knowing the status of everything as it is at this
moment will allow us to* **predict** *the status of
everything as it will be a moment from now.*

Grasping the meaning of those seemingly contra-
dictory statements is crucial to the understanding
of the absoluteness of this treatise; partially be-

cause determinism deniers use the fact that not everything is predictable, to falsely claim that not everything is caused, and therefore causal determinism is not true; and partially because those statements seem, at first reading, to be counter-intuitive. Hopefully, the following will give the reader a clearer perspective:

Whether or not we can look ahead and un-equivocally predict the state of a specific moment in the future depends on the complexity and randomness of events, beginning with the moment in which the prediction is made, and continuing to the specific moment that is to be predicted; and those events may include random and unpredictable events that occur at the quantum level of matter, the level that deals with atoms and particles, (a world of uncertainties and probabilities that we will review later), and we may not know when and what, will cause each of those moments to be what they will be. But, after the fact, while looking back at a moment's preceding moments, though we may not know the mechanics of some of those quantum causes, we can—unequivocally—look back, moment by moment, still frame by still frame, as in a movie, and clearly see that all the moments involved in that chain, **were all determined** to be what they were by all the moments that preceded them, **regardless of what may have been predicted.**

When we look at ourselves in a mirror, we can safely predict with certainty—because it's a simply deduced fact where quantum considerations don't play a part—that the person we see there, will, one day, die; but at what exact moment in history that person will die, would be, because of the complexity of life's moments, and the large number of possible moments till that exact moment, impossible to be predicted; yet, determined it would be, because it will have been caused by all the regular and random moments that preceeded that final moment. And while predictability of a moment is often desireable, **predictability is not a factor in proving either the truth or untruth of this treatise; in fact it is not relevant.**

And, while we are still on this subject of the

meaning of words, what exactly, you may wonder, does a **moment** refer to in this treatise?

In terms of **time** —in terms of how much time a
moment describes— it actually doesn't matter
how you think about it, and, unless you are
about to light a 3-moment fuse on an explosive
device you are holding in your hand, you can
consider a moment to be any arbitrary amount
of time that you want it to be. But the best way
to think of a moment, in this treatise, would be
to think of it as occupying one second of time,
and imagine that second lasting as long as it
takes you to say "one second" at
conversational speed.

As to how much physical **space** we should be
aware of when visualizing a moment? We could
be thinking about a space the size of a particle,
or an atom, or a golf ball, or a basketball, or a
galaxy, or of everything —everything, as before,
meaning everything. Again, it doesn't much matter, the tiny truth of this treatise would apply no
matter the size of the space we've conceptual-

ized in our mind, or that which is required by what we are doing at specific moments. But, generally speaking, the best way to think of how much space a moment should encompass, would be to think of a moment as encompassing all the physical *space* that would, or might, have an immediate, or near-immediate, effect on you prior to, and during, and after, that moment. But the question of exactly how immediate something is, may not be easy to specify. Generally speaking, that physical space would obviously be larger if you are outdoors than it would be if you are indoors; and, in either case, it could change at any time. For instance, imagine that one sunny autumn day you are outdoors in a local park absorbing the beauty of the fall foliage while standing in the shade of a large oak tree, when a dozen feet from where you are standing, you hear a scuffing sound and then, almost immediately, you hear an acorn hit the ground. Now, that may have no particular effect on you at that moment because you might not be paying it much attention, (though, it could be said, that at a minimum, your subconscious mind should be noting its existence);

but, a moment or so later, while you are still standing in the same spot in the shade of that oak tree, you suddenly become aware of that same scuffing sound as before, but now it is directly overhead; and just as you look up to see the cause of it, the same apparently clumsy squirrel drops another acorn which falls and hits you squarely on your left eyeball, and something that a moment or two earlier might have been at the near immediate periphery of your awareness, is now —a painful moment too late— in your immediate area of awareness. So, as you live your life, how large an area should you consider as possibly having an affect on you? Well, as I just suggested, that may physically change from moment to moment, as may your degree of awareness of the change; but, basically, it would depend, mostly, on what you've judged, based on your life experiences, that you should be aware of in a particular circumstance. But, having said that, you should be aware, for the purposes of this treatise only, that in actuality we are talking about everything that can affect us, no matter how distant in time and how large the space, all of which, when we come down to it, means the entire universe and all of historic time. The truth is that everything as it is at this moment was caused by everything as it was in all the prior moments of existence; so, what may actually have an affect on us in a given moment is every thing and every past moment, with the most recent and immediate moments, generally speaking, having the greatest impact on what our next moment will be like; and the least recent and least immediate having the least impact, or perhaps, no relevant impact.

Speaking of moments, and of the amount of space and time they refer to, wakens me to the question of eternity and the eternal, and of a tennis acquaintance who had read the introductory Pocket Edition of this treatise, and had asked me, since he was counting on going to heaven when he died (or so he said, though he may have been pulling my leg), how the tiny truth that governs everything would apply there in heaven. "Would everything as it is in heaven, at this moment, cause everything as it will be in the next moment, like you say it does here on earth?" he asked. Well, I told him I had never contemplated going to heaven, so hadn't given much thought to that question, but the statement that everything as it is at a given moment would be caused by everything as it was in the preceding moments, would seem, to me, to be just as true there, as it is here. "But," I told him, "it appears to me that there may be some other related things that might have to be different there than they are here on Earth, or as they might be on any other planet whose sentient citizens one might encounter there in heaven, and that might possibly include, in some way, or another, our tiny truth. For example, considering that heaven, as I understand it, is supposed to be eternal, perhaps ways of taking the measure of time and space, would be different there than it would be in the factual physical universe we live in —we can't just say, regarding heaven, that something is five eternities long in time or space— measurements of time or space would have to be equated, there, to something, such as our day —which is one complete spin on Earth's axis— is with us, and our year —which is how long it takes our planet to circle our sun. Other planets occupied by sentient beings would have their own various ways of defining time; but what is there in heaven that can be used as a means of setting up a system of measures? We can't know, for instance, if there is any way in heaven to describe a moment. That is something we don't know at this point in our history, something we won't know till some intrepid, scientific explorer, goes there and returns with that answer. I surprised myself that I came up with that response since that question was not actually asked by my friend. But, in any case, he didn't seem to understand what I was getting at, so I tried a different response to that unasked question: "Let us say, for example, that upon applying for admission to heaven, you were accused, by your god, of having committed a vicious crime during your stay on Earth —a crime that you claimed you committed in self-defense— and you were approached by the heavenly prosecutor that was dealing with your case—a very angelic lady— and were told by her, that if you stood trial before a heavenly tribunal and were found guilty, you would be sentenced to eternity in hell; but, if you pled guilty—thereby avoiding a trial— your penalty of eternity in hell,would be reduced to half an eternity. Would you take that plea deal and forgo the trial where there was a very small, but real, chance, that you might be found not guilty, and be free to roam the heavens for eternity?" He didn't answer. I think he thought it was a trick question; so I gave up trying to answer it.

Since I don't have a definitive answer to that unasked question, lets get back to the meanings of words in this treatise: Some skeptical readers may want to claim that **caused** does not always mean **caused**; that the words, **chance** and **random**, may be indicators of events that are **not caused**; and, if so, our tiny truth would not be universally true. But, contrary to what those readers may choose to believe, the words, **chance** and **random,** indicate actions without **expected,** or **regularly occurring**, causes; not actions without a cause. Consider, for instance, this hypothetical **chance** event: someone tells you how they bumped into a friend in a foreign city clear across the globe. They think the odds of that being caused simply by chance have to be so implausibly high that its happening has to be thought of as a miracle magically caused by their god. Of course, if that is true, that god would be the cause. In any case, its happening should not be considered miraculous, no matter whether the word is used in a theistic or non theistic sense. If you search your memories you will recall large and small incidents like that in your own life. I, myself, have had a few of these so-called miraculous incidents.

Once, many years ago, for instance, when I was in the business of demonstrating to the public the attributes of my clients' products or serv-ices, I called in a contractor (whose name I plucked out of the yellow pages) to come to my offices to break through a wall and install a door. He came early on a Friday morning and with a crew of two helpers installed the door in a matter of a few hours. When he was done, I paid him and thanked him. Late that afternoon, my family and I flew to Florida to enjoy a vacation at Disney World. The next morning, after breakfast, as we were working our way through the overflowing crowds on The Magic Kingdom's Main Street, a man suddenly stepped in front of me, and with a look of surprise and disbelief on his face, said, "Hey, Mr. Mel! remember me?" Well, I guess I don't have to tell you he was the contractor that I had met in New York City for the first time the day before; and now here he was, standing in front of me in Disney World. After a handshake, and the exchange of a few words, we both rushed off to catch up to our respective families, lost ourselves among the crowds, and never saw each other again. A mathematician could figure out the odds of that happening to all of us sometime in our lives; but whether caused miraculously by a god, or by simple chance, it would have been caused by everything as it was in all the moments before the event. If those moments had not included our meeting of the day before in New York, that man and I would have passed each other on the crowded street as if he and I were just two of the many faces in the crowd.

As to **random**, most of us would think of a random event as somewhat similar to a chance event in that random events can also be unexpected and unpredictable; but they are different in that they can, at times, be expected, and they seem to occur much more often, though we seldom notice them unless they have a direct impact on us. Also, the complexity of some quantum random events, and their causes, occasionally appear to be inexplicable, and that may cause some skeptical readers to say, if they haven't already done so, that those random events are not determined, that they are not caused by everything as it was the moment before. Those same skeptics may take it further and claim that with all the chaos that exists in the universe, including all the moment to moment interactions of all its tiny, uncountable, atoms and particles —along with its massive planets and stars and moons and asteroids, with their apparently inevitable, randomly appearing attractions and repulsions, and their exchanges of matter, each with the other— how can we claim that everything as it is at this moment, which we are usually incapable of defining fully, was caused by everything as it was a moment ago, the state of which we, sometimes, also cannot define with much certainty? But the search for definitions is precisely what science —with it's search for truth and its goal of creating order out of chaos— is about. Scientists could not have proceeded in their quest for knowledge of our selves and our world without first having grasped that everything in the world, as it is at this moment, was causally determined by everything as it was a moment ago, and that it will cause everything to be as it will be a moment from now. An example of the applied use of that knowledge, would be in man-conceived manufacturing processes, such as on a factory production line, or as with a printing press that at every new moment spits out a printed sheet of paper exactly as planned, exactly as determined by the experienced choices of the printing-press operator, and the generally predictable actions of the press. Of course, presses break down, and operators can make mistakes, and, if so, what once could have been said to be predictable, would then be chaotic and randomly unpredictable; but whatever the outcome, every moment of the process will have been determined by all the moments that preceded it.

Before we get entirely away from the subject of unpredictable random causes; those causes are, by their very nature, the ones most difficult, and sometimes seemingly impossible, to get to the root of (I'll be discussing one or two of those as we go along), their very nature makes them, to those readers sceptical of science, the weak links in the argument for causality; but in reality they are the most obvious and causative links in the argument: regularly occurring moments do very little in the way of causing other than other regularly occurring moments, ad infinitum; but it is the very nature of random events that they often cause moments that are very different than their preceding moment or moments; and they are the cause, for example, of the evolutionary genetic changes that occur in all living things on our planet; and those living things includes us, the homo sapiens of today—the ones who search for the causes of things— and it includes all of our ancestors, including those most ancient, those who scooted from living in sunshine, to living in shade, and back to sunshine again; and from living in water to living on land, and then back to water again, in their regularly and randomly occurring struggles for survival; till a random, genetic-mutation-moment, made it possible for one of them to survive on land alone if it so chose; where began a long chain of occasionally occurring moments, random genetic events that led, over a span of uncountable trillions upon trillions of moments, to us as we are today, the most advanced, though far from perfect, organic thinking machine on Earth.

Here are a couple of other words that need clarifying: **spontaneous** and **accidental.** Those are words that some may think indicate events that are not caused. But the words, **spontaneous** and **accidental**, simply indicate something that happened that **wasn't planned**; they don't mean something that wasn't caused.

And, lastly, to round off this group of words that might be considered as disparaging of the tiny truth that governs everything, let's consider the words, *luck*, and the hypothetical, *if*. The word, *luck*, according to Webster's Dictionary, means "The circumstances that work for or against an individual." The circumstances (in the context of this treatise) being everything as it is at any given moment; all of which would have been caused by everything as it was in the trillions of trillions of trillions of preceding moments. As to the hypothetical *if*, there is no room in this treatise for an *if* that says, "*If* only I had done this instead of that, things would have worked out better." Well, they might have worked out better, we can say so hypothetically; but, in reality it could not happen, there could be no *if*, we could not have chosen differently, we were what we were and the circumstances were what they were; and were we to recreate that event, to the finest detail —including our ignorance of prior outcomes— a million times or more, we would make the same choice every time. (I'll clarify that more fully later on.)

(When I say everything as it is at this moment was caused by everything as it was many trillions of moments ago, I don't mean to say that it was planned by some ethereal being, or beings, as some may believe —this is not a treatise built on beliefs; it is a treatise based on scientific facts and logic that constitute a truth— I'm simply saying that everything as it is at this moment was causally determined by everything as it was a moment ago; and that, if there is a god, anything attributable to it, at any given moment, was caused by everything as it was with him, or her, in the moments that preceded it. For example, that god's decision to create us, if indeed it did so, would have been caused by the state of its mind the moment before it did so; that god could not have created us and then decide to do so; or that god could have created us by accident, or on a whim; which would have been caused by everything leading to that accident or that unconsidered whim, or, perhaps, simply because that is what gods automatically do, or do as a result of a mental orgasm.)

Before we go any further, let me state, categorically, that though there are skeptics who claim that there are events, at the quantum level of matter, that have no cause —that those events occur with no discernable, therefore non-existent, cause— there are no skeptics who could actually prove there is no cause, anymore than anyone could prove the non-existence of a particular god; the only difference being, that while they also couldn't, if they wanted to, prove scientifically, the existence of that god, science can prove the physical existence of the matter and the trigger of the events that those skeptics claim have no cause.

So, having explained the meaning of those words that may be used in an attempt to denigrate the tiny truth of this treatise, here are a few thoughts that may help in grasping what this tiny truth might look like in our daily lives:

Your choice of a mate was irrevocably determined before mating, as we know it, even existed. You were what you were, and your mate was who they were, and neither of you could have chosen differently; a confluence of uncountable trillions of trillions of events (many of them random) that began way back in the mist of time, brought you together; and who each of you were, at the moment you met, sealed the union. Had you crossed each other's path at a different moment in time, you may not even have taken note of each other.

If you say a prayer for the life of a loved one who is ill, and that person survives their illness, it doesn't matter whether you think that your god saved your loved one, or you think the doctors, nurses or medicines saved your loved one, or that your loved one's immune system did it, the cure was caused by everything as it was long before that person was ill.

On a clear night, look up at the moon as it circles our planet on its delineated orbit. If you were told that its exact position, at that precise moment, was caused by everything as it was at the beginning of our universe, you could maybe accept that possibility. But if you were to look up on a sunny day and see a little yellow butterfly fluttering by erratically, and were told: that butterfly and its exact position, at that precise moment, was caused by everything as it was in all the moments in the long chain of moments leading back and back through the haze of evolutionary time to the moment when the first precursor of the butterfly realized it could fly, you would be loath to acknowledge it, you would think it mad to think it so. Yet true it is, and there is no plausible way it cannot be.

Having said all that, here are a few disclaimers of sorts: First, the tiny truth discussed in this treatise is not a new insight. Secondly, it has some naysayers. And lastly, contrary to my assertion that everything is caused by everything as it was the moment before, it seems possible that there may have been one moment when it wasn't.

For thousands of years, virtually all scientists,

and philosophers, and some theocrats, have known the essentials of the tiny truth of this treatise. In ancient Arabia, early-Islam theocrats planted the seeds for most of their current divisive, violent, intrafaith conflicts, when, in the early centuries after their prophet Mohammed's death, they failed to resolve disagreements on the question of determinism and its relation to the line-of-descent of its leaders; while, further west in those turbulent centuries, two early Greek philosophers, Democritus and Leucippus, had proposed forms of determinism that were somewhat similar to the one described in these pages, determinisms that were embraced by most premodern scientists and philosophers of that time, both professional and amateur. The truth described in this treatise, however, is a simpler, but more encompassing, form of those: an absolute determinism that includes everything from the tiniest cell in your body, to the largest acts of creation and destruction laid claim to by the gods, a form that most pre-enlightenment, and some enlightenment, and post-enlightenment, scientists and philosophers would not have dared, and perhaps will still not dare, to contemplate. Today, there are Islamic theocrats worldwide, who, based on their unresolved line-of-descent issues, still wrestle with the question of determinism. Meanwhile, though most contemporary philosophers and scientists agree with the basic tenet of determinism; there are some who dismiss any form of it; especially those forms that claim to be absolute; and they profess to do so based on rational, reasoned, logical, scientific grounds.

A number of those philosophers —along with other interested, knowledgeable parties— appear to oppose determinism on pragmatic grounds. They evidently are concerned that those of us who are less enlightened than they, would not be capable of assimilating this tiny truth of determinism in a constructive way; that they feel determinism should not be thought true because they believe it would be detrimental to our laws and morals to think it so; and they write intricately reasoned volumes asserting some logically deduced fault in it. (I will return to their subjective arguments later, at a more appropriate place in the narrative). The more objective, more technically complex and difficult to grasp, negative arguments advanced by some scientists, and writers of popular science, are based on chaos theory, on Einstein's theories of relativity, and on quantum physics, and are impossible to present and respond to specifically and equitably in this simple little treatise. The arguments stemming from chaos theory, and Einstein's theories of relativity, can be readily researched and dismissed (as they relate to this treatise) by anyone who is willing to devote a bit of time in exploring those subjects; or, if you don't have the time to do that just now, you can just, temporarily, take my word for it that relativity has no bearing on causality, or determinism —nor on the question of free will—a subject we'll get to a bit further along. As to chaos theory: to put it simply, chaos theory states that everything is chaotic at some level, and is therefore random and cannot be considered to be deterministic because it is unpredictable. But, as I've stated earlier, the deter-

minism of this treatise does not require predictability. it only lays claim to the fact that
everything as it is at this moment will determine, everything as it will be a moment from
now, and that it was caused to be what it is at
this moment by everything as it was the moment before, regardless of whether the cause
was predictable; or whether, or not, we fully
understand the cause.

The arguments stemming from quantum
physics, however—with its Copenhagen Inter-
pretations, its Werner Heisenberg uncertainty
principle, its EPR arguments, its John Bell theo-
rem, the questions of probability and entangle-
ment, and numerous other perceptions,
including the effect an observer has on an atom
or particle they are observing (again, as that re-
lates to this treatise)—would be, for most non-
scientific readers, much too difficult to assimilate
judiciously, since, for example, what it means to
observe, is open to interpretation; so I will dismiss those arguments on the simple grounds
that though quantum mechanics is held to be

very accurate in its description of the submicroscopic world of atoms and subatomic particles, some actions of those atoms and particles can only be pinned down on the basis of probability and not on the basis of the certainty that some might claim our tiny truth requires; but when we put together huge numbers of atoms — some fifty million, million, million in just one grain of sand for instance— the probabilities average out consistently and become the predictable actuality that operates at the macro level of our everyday world of thought and speech, and peas and carrots, and knives and forks, and telephones and televisions; and, of more relevance to this treatise, the skyscrapers and flying machines made of glass and metal and rivets and grommets and struts and beams; all engineered using, not the unpredictable quantum physics of the micro world, but the predictable, classic physics of the macro world. If this were not so, no scientist or philosopher, would dare to go up in any of those skyscrapers or flying machines; nor should we.

There it was, again, the issue of unpredictability because of quantum mechanics. The quantum arguments against determinism are mostly based on the fact that there are events in the quantum world that cannot be predicted other than as a probability; and, perhaps more relevant, that there are scientifically examined events that can't be actually predicted at all. Take the case of a chunk of radioactive uranium. It is impossible to predict when a specific atom —of the uncountable trillions in this one chunk of radioactive material— will decay by emitting an alpha particle; and therefore that decay has to be considered to be random and unpredictable; but, to reiterate again: predictability is not a factor in the tiny truth of this treatise. However, there is a more serious contention by those who argue against determinism, the contention that because the cause of a specific atom's decay at a specific moment of time cannot be identified, there is no cause —they insist that nothing whatsoever can be the cause of that decay at that specific moment —and therefore causal determinism is not true. But, as I've stated earlier, the amount of atoms that will decay in a given, macro time frame, are consistent. and predictable.

Lastly, on a larger scale, it has to be admitted, when discussing causality, to the possibility that not everything is caused; that there may have been a moment in time, in the history of everything, that had no cause because there was no need for a cause, that the moment of the beginning of everything had no discernable cause because it is quite possible that there never was a beginning, that the matter of the universe always existed, that it existed —as, I understand it has been proposed by renowned physicist, and cosmologist, Stephen Hawking— in the form of a singularity containing compressed matter and energy of infinitely high density and temperature which, at a point in time, expanded explosively to create the universe as we know it today. I think we can accept, with reasonable scientific certainty, that the sudden, explosive expansion of the matter and energy in that singularity was the cause of the birth of our known universe; but where that initial matter and energy came from, and what caused it to be, is not unequivocally known, and is an open question; though many say that matter always existed and is currently in a state of explosive expansion that will be followed by a state of contraction, followed again by a state of explosive expansion, ad-infinitum. Others ignore that possibility; and some religions —the ones that accept that there was a Big Bang— say it was their god that created the matter that exploded and became our expanding universe as it is today. But that brings up the follow-up question of what caused that god itself, to be? Perhaps, those who believe, will argue that their god, or gods, always existed, and therefore there was no need for a cause; or they may argue that their god, who didn't exist, created itself from nothing, and then created the universe from nothing. Well, generally speaking, each of those claims may appeal to some segment of our human population; but, rationally speaking, it would be less of a stretch, and more in keeping with the simple scientific nature of this treatise, to just state that the basic, fundamental nature of matter, has always simply been

to be, to simply exist rather than to not exist;
because if matter's nature was to not be, to not
exist, it would have required the arguable com-
plication of considering some sort of a supernatural force, such as a self-created superman,
or superwoman, or god, for us to exist and be
here to consider it. So, with that caveat to the
use of the word, everything —and in the interest of keeping things simple rather than getting
into what might become the necessity of exploring, without end, and without hope of dis-
cerning it, the abstruse role of enigmatic
antimatter— I'll just keep stating that everything
is determined by everything as it was the moment before; and, mean by it, only the moments
whose existence we can reasonably account for
scientifically, meaning all the moments since the
explosive event known as The Big Bang.

All this about the beginning of everything, and
about quantum theory, and about singularities,
brings up, once again, the question of **predictability**: Nowhere in this treatise is it claimed
that because everything in our everyday macro

world, at this moment, was caused by every-thing as it was in all the trillions of trillions of moments that preceded it, everything is pre-dictable. That is not to say, as I've also stated be-fore, that short term, simple things are not predictable. For instance: you are standing in the middle of a desert; you have a cannon ball in your hand and you let it go; we can safely pre-dict that the ball will fall to the ground, we all know the law of physics called gravity, we know it because we experience it every time we drop something, or every time we fall. And while there are many other things that are sim-ple and predictable, there are many more that, though causally determined, are, like the follow-ing examples, much too complex to predict:

Get a large barrel and fill it to the brim with dice; take the barrel to the top of a high hill; dump all the dice down the hill's stone face; watch them roll down, tumbling, bouncing off each other and the hard rough surface of the hill; watch as they come to a halt at the flat ground at the bottom; look at the numbers that are face up; they would all have been determined long before dice were thought of, but no gambler or math wiz would have dared to attempt to predict them; nor could anyone replicate, exactly, that roll of the dice. But, if a movie had been made of the event, and we played it one frame at a time —from the end to the beginning, or the beginning to the end— we could see clearly, frame by frame, exactly what caused each dice to be in its final position.

Think of a maple tree in the autumn, its leaves all oranges and yellows and reds. Exactly when each of its individual leaves will fall to the ground —the random fall of each leaf dependent on a multitude of factors, including, but not limited to, the actions of moon and sun and rain and wind, and the whims of squirrels and birds and bugs and bears— would be impossible to predict. Ask any arborist, or physicist, if they could predict even one. Yet, the which and where and when of their fall was determined long before the tree sprouted from the ground.

Perhaps the thing most difficult for us to predict correctly —and the one we are most familiar with— is the predicting of the weather. If you were to stand outside at the corner of Elm and Clover at eleven a.m. on a sunny spring day —a clear day but for the few clouds that hovered at the horizon— and you took out your smart phone and asked it what the weather would be at that corner at exactly three p.m. that after-noon, you could probably receive a decent ap-proximation of it; but I have no doubt that you would not bet everything you own on its being exactly, to the nth degree, what your smart phone told you. And, if the prediction given you is for rain, you can be sure that your smart phone will not be smart enough to predict, if you asked it to, exactly how many rain drops will fall on a specific square foot area of that corner beginning at three p.m. and lasting for a specific amount of time; yet it will all have been caused by all the moments preceding that three p.m. moment and all the specified moments that followed. The results of the preceding, complex examples, would be, if tried, impossible to replicate and test on later attempts; and not just because of the extreme difficulty, and in some cases such as the weather, the impossibility, of physically duplicating everything exactly, as it had been at the moments of the original attempts, but also because of the impossibility of being able to attempt them at the necessary same moments in historic time as the original attempts. Every specific moment, in its universal entirety, exists just once, it causes the next moment, and then it is gone, it is history and cannot be physically resurrected and lived again to create a following moment. With that in mind, I wonder what the heaven, that the acquaintance I mentioned earlier aspires to, is like. I don't think that anyone who is reading this, and might be on the verge of a blissful eternity in heaven, will really worry about what it would be like, as long as they can count on their stay there being eternal; which, to me, is the most difficult aspect of heaven to wrap one's questing mind around. I, for instance, can't quite imagine what it would feel like to be there day after day for a million years, which is a tiny amount compared to being there for eternity. While you're in the act of mulling that thought about eternity over in your mind, think about the brain where that thinking is going on —a brain that contains some one hundred billion neuron cells in which are stored the totality of your mind and its knowledge, the totality of the mental you. How did those billions of individual neurons, with their connecting synapses, learn and evolve and adapt, from the moment you were born, as they were exposed to the world they were born into? Could anyone who knew you at birth have predicted exactly when your mind would figure out that one plus one always equals two, and not the three that some may, perhaps, have maliciously indoctrinated you to believe? At what moment you figured it out, if you ever did, would have been determined long before man figured out they could stand on two legs.

So, now that I've acquainted you, as best I can in this little treatise, with the tiny truth that governs everything; how, you may ask, would that knowledge change how we view ourselves and the world? And how would that

make us and the world better?

Well, to put it bluntly, and to get quickly to the crux of the matter, our little truth means **we are not self-made,** it means we did not create the we that we are at this moment; it means we were created by everything as it was in all the moments that preceded this moment, all the way back those trillions of trillions of trillions of moments ago; it means that we were created by a god or/and every one of the very long list of our direct ancestors, and by all that this "we" that is us has been taught and absorbed, and experienced and learned from birth till now; and the exact person that we are at this moment —to the tiniest detail— **is the only person we can be**. And that means **we do not have free will** —that is, we are not capable of making choices other than the ones we make— it means we have will, and it's ours, but **it is not free to choose other than what it chooses**. It may feel to us, as we live our lives, as if we *are* making choices between two or more options. And we are, but the actual choice we make at any given moment **is the only one we can make at that moment.**

There are experts, in the neurosciences, who say that the brain will sometimes make random choices when the choice to be made does not present any obvious good options; they say that in those potentially critical moments a random choice is the fittest choice in terms of evolution and its gospel of survival of the fittest, where a random choice is better for survival than no choice at all; so an automatic mechanism for causing random choices not governed by our conscious mind, must have evolved very early in the evolution of our brains. If that is true, as it seems to be, it must also be true in the evolution of all animals —actually in all sentient beings— and, if so, it means randomness plays a very important role in that evolution, and not just in the random fusing of our DNA with that of our opposite-sex partner. We humans, masters of our world, don't want to know that we can't make free choices, that the next choice we make will not be of our own making, not made by the free-thinking "me" that we think we are, that it will have been made because of a long chain of past events beginning way, way, back, and leading to the moment we make the choice. We feel that to admit that — **that we don't have free will**— means we would be admitting that we are nothing more than mindless automatons; that we'll obey who we're told to obey; that we'll believe everything we're told; that we'll do anything we're told; that we'll love who we're told and hate who we're told; that we'll believe the most childish fantasies; that if we have no free will, no will of our own making, we will be nothing more than placid, malleable sheep induced to follow any Pied Piper who promises to lead us to green grass and cool water. And we would be right, all those things can, and have, happened; **but they don't happen because we don't have free will, they happen because we think we do.** Allowing ourselves to think we do is how the Pied Pipers of the world, including those we may revere as gods' ministers, get us to convince ourselves that we follow their beguiling musical tweets of our own free will, and we feel we can turn away anytime we want to. Most of us believe we are free to think whatever is thinkable, and choose whatever is chooseable; and we believe it because we think there is a separate, inviolate, autonomous "me" that resides in us, a distinct "me" that is there from the moment we're conceived, a "me" that is virginal and pure and not influenced by its creator—be its creator a group of genes or a god or both— an individual "me" that's able, from the moment we're born (or perhaps before), to use its supposed free will to make choices, to discern whether incoming information is factual truth or spurious fiction or something between. It is, however, difficult to understand on what basis we could make such free choices since our virginal minds would be virtually empty of any sort of input or knowledge on which we could base our choices; and if it is not virginal, it must have been impregnated by a god and/or our genes, with some at least rudimentary knowledge; and therefore the choices we make would be based on what knowledge we have been impregnated with, and are not ours to will freely, but are dictated by that god and/or our ancestral genes. But, there are others of us who believe that except for our genes —which determine our physique, the overall construct of our brain, and, possibly, some, of the rudiments of our mind— there is no autonomous "me" when we are conceived, only the gradual emergence of a conscious awareness of ourselves as we develop, first in our mother's womb, and then out in her arms, and, finally, in the broader arms of mother Earth. And, then, there are those of us who believe that we are born as naked in mind as we are naked in body, with no sense of anything at all till we gradually learn, by osmosis, from the world we are born into. But it doesn't matter whether we think of our "me" as our "soul" or just our "me;" or whether we believe we are born with a blank brain, or with a brain already having somehow begun to process and use incoming information to form a mind; or, as recently confirmed after years of research —using functional magnetic resonance imaging (fMRI)

and psychological testing— that certain elements of our brain —the prefrontal cortex, and
the amygdala, and, to be precise, their interactions, one with the other— can cause us to start
reacting in specific ways, at a very early age, to
the immediate world that surrounds us; and,
depending on the neurobiological construct of
those elements, to do so in very distinct ways,
with different degrees of empathy, and with
varying perceptions of fairness and justice and
other related traits, that for many of us, tend to
result in perspectives that result in making
judgements and choices. on a continuum of
multifaceted greys. Others of us, on the other
hand, present strong traits of loyalty to our family, race, and nation, and a high degree of re-
spect for authority and hierarchy and sanctity,
and discomfort with uncertainty in general, and
with the new and untried in particular, and
seem to have fixed black and white opinions on
the questions we all seek answers to. These
traits, according to the researchers, stay with us
stronger and longer —usually for our entire
lives— the more they are derived genetically

rather than from indoctrination; and these are traits which, according to the neuroscientists working on this research, strongly correlate to our philosophical and political stances —whether they be liberal or conservative— stances so embedded in us that liberals and conservatives, wherever they live in the world, cannot grasp, to any sensible degree, why their opposites believe as they do; which makes it virtually impossible for one to convert the other to their views. But, regardless of whether we accept any of the above scenarios, or some amalgam of them; we need —in addition to those traits we've inherited genetically— someone with experience of the world to be our guide after we are born, someone to inject our spongelike brain with contemporaneous knowledge that will teach us how to learn, how to understand and survive, in the specific world we are born into. And that someone is, generally, our parents and also grandparents and older siblings, and aunts and uncles and cousins, and anyone else who is active in our lives; and it is they who initially teach us about that world, and they teach us according to their beliefs and under-standings; and, since in those earliest years, there is no one in our lives, as powerful as they, who can contradict what they are teaching us, (and add to that the powerful conjunction of our brains with those of our blood relatives, whose brains are to some degree a close ge-netic mirror of our own), means that their ideas, their view of the world, their philosophy of life, become imprinted easily and indelibly, without recourse, in our minds. Then, as we grow and we come in contact with the wider world, our imprinting is strengthened, or, to some degree, modified, by our teachers and our playmates, and, later by books and magazines and newspa-pers, and music, movies, television and the inter-net. And, after all that —when we are finally adults—the "we" that we are is the product of all of that; we are the product of our biology in general, and, more specifically —if you accept the latest scientific research— the biological construct of our brains, and all the compound-ing, or modifying of the imprinting we've re-ceived from our nurturers; and unless we have

been taught to keep our mind open and critically questioning, that imprinting becomes set,
like concrete, in the fixed, but possibly malleable, mold that is our brain. But regardless of
whether our minds are set like concrete, or our
minds are still pliable, what we are at a given
moment is what we are, and we cannot be otherwise, and what we do at that moment cannot
be other than what we do. But when we *think*
there is an inviolate 'me' that is not ruled by
what we've inherited —and what is perhaps to
some degree as apropos here, what we've been
imprinted with and experienced— we don't
question what we are doing, we don't question
the brain we inherited, the indoctrination we've
received as children and the imprinting we've
continued to receive throughout our lives; and
that may cause us to do things that are not in
our best interest, or in the best interest of
those we care about, or in the best interest of
the society we live in, or that of the world at
large (witness the indoctrinated, dogmatic ter-
rorists who are inflicting their horrors on the
world, as I write this in 2016). And that is the

danger. The danger is not the fact that we don't have free will, but that we think we do; and that because **we think we do**, we don't develop the tools of critical thinking and questioning that would allow us to make informed and examined choices instead of the sheep-like, herd-like, choices that dangerous, dogmatic indoctrination leads us to make. And the worst danger of all is that, because we believe we have free will, we indoctrinate our children in that believe and we don't teach them to bring an open but critically questioning mind to the consideration of new ideas, such as the very one elucidated in this treatise. Unfortunately, that leaves them helplessly stuck, for a lifetime, in a bog of antiquated, narrow beliefs, or captive to a whirl of new fads foisted on their unquestioning minds by promoters of fashionable products or philosophies. And for those of us who think we have free will because a god gave us our soul, and, with it, free will, it must be that when that god speaks of free will, she or he simply means that he or she gave us the ability to make choices; because, if that god knows all, she or he knows that we can make choices, although we may make them subconsciously (as Daniel Wegner, and Sam Harris, and Mark Twain, state in their respective books, *The Illusion of Conscious Will, Free Will,* and *What is Man*), but that the only choices we can make, consciously or subconsciously, are the ones we make; and that is not the free will that most of us would like to believe we have.

The "we" that we are at the moment we make the choice, and the event that requires the choice to be made, determines the choice, and that choice is the only one we can make. Our brain, with its mind, or soul, is the only tool we have that we can use to make those choices, and it can only be as it was originally shaped by our genes, and later, as it was codified, or modified and remolded at the hands of our parents, and then fiddled with by anyone else that had access to us by word or example. We were at the mercy of all of that, we had no control over who we were to become.

(That does not mean, however, that we are all the same, as if we came off an assembly line, as a robot might; it does not mean that we don't have a will that's our own. Quite the contrary; we, all seven billion plus of us, are all gloriously unique: we are all, including identical twins, different at birth, and, except possibly for twins, at significantly different moments in our family's and world's history; we may live in different areas of the world, speak different languages, have different accents— and have different belief systems, and be in different socio-economic and political environments, and had different experiences, and been indoctrinated differently; all of which possible combinations, if totaled up, would add up into trillions upon trillions upon trillions of trillions of different possible configurations, making it quite impossible for there to be, or to ever have been, or ever will be, two of us humans exactly alike, with identical wills).

As to robots coming off an assembly line: When we humans begin to manufacture robots that are human-like —and possessing what their creators will refer to as artificial intelligence— the first of those humebots will be, unlike we humans, identical to each other, just as if they came off a factory assembly line. They will march in lock step out of the factories that produced them, each one looking and behaving indistinguishable from the other, and destined for the menial labors for which they were programmed. But the developers of those first robots won't stop there. After those first-generation robots have proven their utility, their creators will make each succeeding generation of humebots more capable than the preceding one, and soon there would be humebots that will be visually, and intellectually, similar to us; they will be capable, if instructed to do so, of ambling out of the front doors of their incubating mills, surveying their surroundings and finding their own way to their assigned destinations.

Eventually there will be generations of advanced humebots that will have fully developed human-like senses and feelings, and they will be programmed to fill ever more demanding roles in our increasingly high-tech societies; and soon after, among other traits, they will have been programmed to confront newly evolved problems ——problems never before encountered—— and, on their own initiative, explore all feasible solutions to those problems, including solutions that may not have occurred to, or been imag- ined by, the most intelligent and educated of us blood and bones humans. And much of that ability will have come from the knowledge and world views they were indoctrinated in by their human, proprietary masters; and, as with us, and other sentient beings, they will learn, by observ- ing their human elders and care givers and mimicking, and mastering, by trial and error ——with its-survival-of-the-fittest mandate—— the ways of the world in which they will be navigat- ing. And just as we are the product of our genes, our environment, and the hopes of our parents, humebots will become the product of those who engineered and produced and pro- gramed them, and of their proprietary master's wishes, whatever those might be, and by what they observed, and learned, from the environ- ment they were created in. But, what will be the future, in the long run, of these increasingly super-intelligent humanoid bots who will be in many ways virtually indistinguishable from us? Will they be treated as our equals? Will they have, in the religious sense, a soul? Will they go to heaven, or hell, when some accident causes their useful life on earth to come to an end? Will some of them develop liberal philosophies and tendencies, as some of us humans possess? And will others develop conservative philosophies and tendencies like others of our species?

Will they develop prejudices like ours? Will they, for instance, look down on the first humebots and think of them as dumbots? And, most importantly —what some developers have a nagging concern about— will their humebots become so intelligent that they will learn how to replicate themselves, and become so ubiquitous that they will begin to contest our power over them? So numerous, that eventually, they will overrun our planet and wrest away control of it from us? Become so powerful that we will eventually become their slaves, as they, it could be said, were once our slaves? And will we, the human species simply become, irrelevant and cease to matter, and join the legions of species that have become extinct — our remains being buried in the vault of some out-of- the-way museum? And will these everlasting humebots, (who will live for eternity because their aging parts will simply be unplugged, and replugged with new, more advanced, parts), become the masters of the entire universe? There doesn't seem to be any reason why they couldn't do so; given that they could each function efficiently for many millions of years—enough time to reach, and occupy, planets well beyond our solar system, well beyond our galaxy.

But how will we be able to counter that possibility, the possibility that the humebots will become our masters, and we their minions? We could, withhold all basic underlying, scientific knowledge from them; knowledge not required for their specific, assigned work. But that may prohibit their ability to solve broad-scoped problems, and seriously hinder their utility; reducing them to being specialists in narrow branches of endeavor. No, we would need to bring some of them to a very high level of overall intelligence that would make them of viable use to us in creative and administrative and leadership positions, but not so intelligent that they will see how ignorant, relative to themselves, we are, how vulnerable we are, how easy it would be for them to make themselves the masters, and we the minions. To avoid that frightening possibility, we would need to change ourselves; we would need to match the humebots, step by step in intelligence; and to accomplish that, we would all need to have genetically engineered brains, and develop them to the maximum utility possible; and, failing that, or in addition to that, make use of expansive, internal brain implants, and externally worn, wireless computers that would massively improve our memory capacities and the speed of our calculations. We could make ourselves a super-intelligent species, in a broad sense; while we would give the humebots only that knowledge necessary for the specific work they have been created for. However, while each humebot would only have the specific knowledge relative to its work, if they got together and integrated all that specific knowledge, that merged knowledge might open the door to the broader knowledge that we humans would have kept to ourselves, knowledge that the humebots would need in order to advance beyond their narrow, specialized status; and we would have to be forever watchful of that danger, and prevent the humebots from acquiring that broader basic knowledge that, if it became known to them, they could build on and become unstoppable.

Here's another possibility: what if, when we begin making intelligent robots, we indoctrinate them to believe what, of course, will not be true: that they have free will —a will that is free to choose whatever it will; but, in reality, they would have a will that will be choosing what we, or their environment, influenced them to choose. If we could convince them that they have free will, we humans could then justify punishing them if they attempted to elevate themselves above their assigned place in society; we could sentence them to a humiliating life as a dumbot, a life of cleaning out toilets and animal pens, and other ignoble, or dangerous, jobs. And we could, in extreme cases, sentence them to humebot death, the extreme act of dissolution. And we could justify doing that by pointing out, to other humebots, and society at large, that those convicted humebots had the free will to do right by obeying their master, but instead chose to disobey them and do wrong.

And we could justify that punishment to the goody-good humans among us who might say the sentence is too harsh, by saying that it is the only way to protect the human population from the potential hubris of egotistical, overreaching humebots who might come to think they are better than we are; who might think they deserve to have those things that we humans have —those things that we've worked so hard for— and who might think they have the right to take those things from us by force or wile. If that should become a real, possible threat to our way of life, it would require that we humans change ourselves; it would require, among other changes, that we make ourselves more pragmatic, less empathetic; and tougher, more willing to fight fire with fire, to do whatever it takes to keep humebots in their place, to destroy even those we might consider as members of our family. But that change won't be easy for many of us of liberal leanings; It would require us to be other than what we are, what we were made by those who made us.

Many of us may not see that as a problem, we may think we can change the "we" that we are; and we can, but, without some external impact on our beliefs it is difficult to see how we could change who we've become, who we've been indoctrinated to be. In order to make changes on our own initiative, we would have to jettison long held beliefs; many of which are ironclad because they were ingrained in us at birth, and then solidified by indoctrination in our early, vulnerable years.

Changing ourselves is difficult even if our minds do not have beliefs and attributes that are set like concrete, and even if we want to change the "we" that we are. Only if we've learned the value of having a critically questioning mind, and

being open minded —of not being dogmatic—
and if we've been exposed to mind-altering
knowledge or experiences, can we change who
we are. Sometimes a new bit of information
that we are unexpectedly exposed to can fill in
the gap between two seemingly unconnected
pieces of knowledge that we already possess,
creating something new in our minds —a new
vision of the world for instance. But sometimes
an experience can dramatically, and forever,
change us and our vision of the world regardless of the state of our mind, our will, or our
wishes. Take the example of a strong young man
who thinks himself heroic: One day he is in a
local grocery when a hooded man, possibly a
humebot, enters the store and suddenly points
a murderous looking assault weapon at everyone in the store and tells them he will kill anyone who moves. As the desperate, wrong-doing
bot tells a clerk to get him all the money from
the cash register, our hero decides —that is, he
chooses, because of who he is at that moment—
that, at the first opportunity, he will leap at the
presumed humebot and wrest away his gun.

Suddenly, at the sound of a police siren, the humebot turns to look out the window, and our hero is about to leap at him when another young human, also envisioning himself a hero, dives at the humebot. But before this new hero can get his hands on him, the humebot quickly turns and fires his assault weapon at him, effectively blowing his head to pieces and killing a neighboring human in the process. Instantly, on witnessing this, our original hero becomes a different person than he was. Some might say he has become a coward; but, more likely, we should say he has been profoundly changed, and not by use of his will but because the experience has made him a different person, a pragmatic realist; and when the killer bot escapes out the back and is never caught, our hero may, in addition, become a cynic—one less able to have faith in society's ability to protect him from violent bots.

Though it may be hard for some to accept; what I've demonstrated about free-will applies to the so-called immaterial world

as much as it does to the material one.

Those readers who may have the most difficulty accepting the truth of this absolute determinism, are those who believe in the existence
of ineffable, transcendental, ethereal things like
goblins and ghouls and ghosts, and spirits and
souls, and gods and angels, and the efficacy of
prayer, and heaven and hell, and in the ability of
some humans —those with extrasensory per-
ception—to see into the future and the past
and to contact the souls of the dead. Those
readers feel certain that those things have no
material aspect, that they are supernatural, and
they can't imagine how the tiny truth of this
treatise could apply to them. And, as to their
god, they will say, "Our God is all powerful, he
can do anything, he has free will." That a god is
all powerful in every other sense may or may
not be true, but our simple little truth is an unequivocal truth that applies to gods as well as
to ourselves, and to souls and ghosts and spirits
and angels and all other things which, if they
exist, appear to have no material aspect. It can

not be otherwise; neither gods, nor angels nor any other so-called immaterial being can do something, and then, only after the act, plan on doing it; they can't scratch their butt and then decide to do so, although they can do so spontaneously as an unconscious reflex to an itch, or an urge —that is, if supernatural beings have such things as itches and urges. The truth is that not even a god can alter the fact that everything as it is at this moment was caused by everything as it was a moment ago —a given moment cannot precede the moment that caused it— but, if a god had the power to change that physical law so that it read, "Everything as it is at this moment will be caused by everything as it will be a moment from now," it would mean that particular god changed the direction of time, and our universe would begin its swift dystopian degeneration back to the moment of its beginning, if in fact there was a beginning; and if there wasn't any, if it always existed, it would simply continue to exist as raw matter for future worlds; worlds inhabited with sentient beings; beings who, like us, would wonder why there was matter, rather than no matter, and wonder if something more utopian than their world may once have existed, and may exist again in the future.

Whether that makes sense, and, if so, how long that dystopian degeneration and utopian regeneration might take, I will leave to those with a mind more deductive than mine. But, if the heavens and the gods and the soul do exist, they are not nothing, they are something. The fact, as claimed, that these things are not analogous to the material world does not mean they are not subject to our tiny truth. And if there is a hell, it also is not nothing, it is something, and it too must observe that truth. And, as I've already stated, neither in heaven, or hell, or on earth —not now, nor in the future— can you, nor will you, scratch your butt before you consciously think about doing it, or you may do it spontaneously, without thought, because it will have become a subconscious act. If this one encompassing truth weren't a fact, the heavens would be incapable of being, and gods' minds would be incapable of thought, and the same would be true of our "soul" or our "me."

At this point, you may be saying to yourself that a lot of this about the one truth may be true, but that, nevertheless, you feel that you have free will and that you can change your mind anytime you want to and do things differently than you might do otherwise. To prove it, one Saturday morning you get up earlier than you normally would, deliberately changing your routine. You take a longer, hotter shower than normal; you get dressed in exercise clothes you haven't worn in years; and you go for a morning walk—something you've never done—in a wooded area that you've never been to. You follow a well-worn path that seems to circle around through the woods and back to where you started; but after you've been walking a while you come to another path—a path, not so well worn, that seems to turn off deeper into the woods—and in order to be different, in order to prove to yourself that you have free will, you take this less traveled path; and sometime later you suddenly, unexpectedly, come upon a big momma bear caring for her two cubs. She turns your way and instantly determines that you are a threat to her cubs and she is on you before you can react, knocking you down, slashing your throat, going for the jugular. Then she leaves you there bleeding to death. And so, you would say, if you were capable of saying anything, "See, I changed what it was determined I would do, and I changed the moment it was determined I would die." But the fact is that you wouldn't have changed anything. The fact that, hypothetically, you had been in a bookstore and were intrigued by an earlier version of this treatise and had begun to read it, is because of who you might have been as a person at that moment in your life: the kind of person that led you to be in the bookstore in the first place, the kind of person who wanted to read that book, and the kind of open-minded person who liked to disprove other people's assumptions; and, because of that, you got up early that morning and did all your morning rituals differently and went for a walk in a place

you had never walked before and took an alter-
nate lightly used path deeper into the unknown
and found yourself face to face with death.
That's the person you were when you ostensibly picked up that book, and you could not
have done anything differently; everything as it
was at the moment of your death would have
been caused by everything as it was the moment before, including every thought and emo-
tion in your brain and the existence of that
book, that path and that bear. And, whether this
were true, or just the story that it is, the moment of your real or fictional death would have
been determined a long time ago, long before
you entered that bookstore, long before
man was man or bear was bear.

The preceding also applies to any god that may
exist. That god might one day choose —in order
to be different, in order to demonstrate that
he, or she, has free will— to terminate our universe and create an entirely new one. If that
god does so, it is because of who that god is at
the moment that god makes that decision, and

that god would be the cause of all the effects of it. That god could, alternately, because of who that god ostensibly is, completely eliminate our universe as if it had never existed, as if it never occupied a slice of time, or a space, in that god's own memory. But if that god did so, that god would then be at the exact same state he or she had been at when she or he first created our universe, and that god would do everything the same as she or he had done it before; creating it exactly as he or she had done it before, and, possibly causing you to be reading these exact words just as you did once before.

At this stage of the narrative, you may have become convinced that you can't change who you are, or the things you do, and you may say to yourself, "If everything that will happen in my future life will have been causally determined by my past life, and be unchangeable, I might as well stop striving to accomplish anything and just sit on my hands and do nothing and wait and see how things play out." Well, if that is who you are and who you

want to continue to be, that is what it is determined you will do, and you will bear the conse-
quences of that. But if, instead, you go on
striving to do something productive with your
life, then that is what it was determined, by all
the prior moments of your life, that you would
do; and the life you will continue to lead will be
a result of that. In either case, you can only act
according to what you are and what your world
is, which was all determined those trillions of
trillions of moments ago.

Some final considerations and conclusions:

"So," you may ask, since I haven't specifically
brought it up yet, "If we don't have free will and
can't make choices other than the ones we
make, choices based on the genes we inherited
—including those that determine the construct
of our analytical, contemplative brains, brains
that will proceed instinctively when contempla-
tion is not possible— and on the indoctrinations and influences that we've had imposed on
us, or that we absorbed, unawares, from the immediate community we grew up in— how are

we responsible if we break the law and harm a fellow member of the very community that failed to raise us to be a caring, empathetic adult, an adult that would be an asset, and not a burden, to it? How is it justified that we be punished for that? Shouldn't that community, itself, have been considered guilty of causing that harm, guilty of failing to raise us to be good citizens who would not do a fellow member any harm, other than in the act, if necessary, of defending one's self, or one's community, from a fellow member gone wrong? Well, they should; but raising everyone to be good citizens is an impossible task if our community does not subscribe to an egalitarian ethic (or, to be more specific, an equalitarian ethic), if it doesn't treat all its members as equal citizens, if it does not provide each of them an equal opportunity at success in life, if it doesn't encourage, and provide the means, to allow each one of us a chance to enhance our abilities according to our innate attributes and desires, and of using that learned proficiency to contribute to the wellness of our community; and if it doesn't encourage us to dream and soar as high as our wings will take us; and if it doesn't encourage entrepreneurship and scientific progress in all fields of endeavor, including basic research; and lastly if it doesn't provide for the health, and especially the mental health, of its members. And all that, while encouraging us to take responsibility for ourselves, our families, our friends, and our neighbors, and only turning to the larger community when the need is beyond the scope of the individual, their family, and their immediate community. And all that, while reminding ourselves over and over again, that we, all seven billion plus, plus, of us, are one grand community, living together on one little-blue-planet; and that we are each responsible, individually and conjoined, for keeping our little slice of the universe livable; for keeping it beneficent, bountiful and beautiful.

But, having said that; successfully raising all of Earth's children to be good members of our community simply by creating a more equalitarian culture, is expecting, in the near term, more than man, individually, and as a community, can deliver; witness how many parents choose to indoctrinate their children in believing in the same ancient deadly dogmas that they themselves were helplessly indoctrinated in; while how few of them help their children to develop an open, but curious and questioning mind about everything they brush up against, including when they are asked to believe in something on just the say so of some imposing adult to whom they are a captive audience. And witness our inability to see that the worldwide availability of free contraceptives; and of safe, free, legal abortions for those single women who are pregnant with an unintentioned, unaffordable, unwanted child—an untenable pregnancy often occurring, in part, because of the very lack of access to those contraceptives—would be, in the long term, an obvious win, win for humanity; but, instead, those beneficial practices are blindly frowned upon; and are even considered to be criminal offenses among many segments of the world's population—even among some of those segments that think of themselves as being among the most progressive and compassionate, but especially in those where their god's representatives forbid those practices, under threat of eternal hell. This blindness to the value of those preventive contraceptives does tremendous long-term, maybe irreducible, maybe fatal, harm to our relatively small planet, a planet with limited carrying capacity, and it does so on two levels: one is that unwanted children —especially those born to young, single, uneducated girls without financial resources— are, as a group, aside from their adding to the overcrowding of our planet, the ones significantly most likely, as teenagers and scarred adults, to perpetrate violent criminal acts; the other is that the current, almost universal, unavailability of these free preventives, is the largest contributing factor in the overcrowding of our small, vulnerable planet; a situation that will inevitably, because of an ever increasing shortage of resources, including food, water and liveable space, contribute to a rise in criminal behavior and the instigation of desperate, brutal, life-or-death, perhaps apocalyptic, wars in serious competition for those diminishing resources. So, till we as a society can bring ourselves to nip in the bud these and other potential causes of destabilizing conflicts, and brutal crimes, we must continue to demonstrate, as a deterrent to it, the punishment of the perpetrators of those crimes against humanity, by forcefully isolating them from society.

Unfortunately, at present, how we, as a society, decide to dole out that punishment, and the language we use in the sentencing of it, is directly attributable to our unmindful attitude towards the less fortunate among us, and especially to those of the less fortunate who will be most affected by a dystopian world; those who, when the world was relatively well, had already turned to a life of crime as a consequence of having grown up in disadvantaged circumstances, including those mentioned above and below. So, those of us who are fortunate enough to have a productive and rewarding life, would do well to remember to remind ourselves that though we are all considered to be equal in the eyes of the law, we are not equal in our ability to abide by the law; and that any combination of personal, mental, physical, educational, or environmental deficiencies, may cause any of us to engage in criminal behavior; and rather than demonizing such acts by automatically terming the persons who commit them as evil and vile and monstrous and depraved and diabolical and unredeemable —ugly words that prosecuting attorneys and the media are fond of promoting; words, among others, that are maliciously meant to incite vengeful passions rather than the intellectual, cool-headedness required of jurors and judges—we should consider that the use of less inflammatory language, along with recognizing, and considering, what caused the perpetrator of a crime to be who they were when they did what they did, could lead to fairer trials with more just and equitable outcomes.

As to the role of religion on this issue of crime and free will, there are those who claim that exposing the truth that we don't have a will that is free to choose other than what it chooses, would, if believed, destroy a religion's ability to use their promise of heaven, and, es-

pecially, their threat of hell, to deter wrongdo-
ing. Whether there is any truth to that claim is
impossible to say, since, in general, many, if not
most, humans have seen fit not to be concerned about any of their gods' promises and
threats; witness all the wars and killings and bru-
talities and crimes and inhumanity perpetrated
man on man, across the ages and around the
globe; and, this, despite the fact that many, if not
most, people who have committed those atroc-
ities —which includes the abomination of slav-
ery— were, or are, professed people-of-God.
And it is quite impossible to assess how many
people-of-God who don't commit those atroci-
ties, deter themselves from doing so out of a
fear of their god's judgement. And if they act
humanely, there is no way for us to know that
they wouldn't be just as humane without fear of
their god; witness the many people who are
not people-of-God, who are humane. After all
—gods or no gods— we homo sapiens, have the
potential of feeling empathy for our fellow man;
we are capable of donning each others' skin
and attempting to understand how each other

feels; we are, after all, kith and kin, we are all descendents of our species' first mother. But none of that matters; when it comes to our opinion of it, we will accept, of it, what our mind—a mind not of our own making—chooses as to whether religions' claims that we possess free will, are valid; that we are able to make a choice other than the one we make, to choose other than what we choose, to choose between doing good or doing bad.

If you have a firm belief in one or the other of the gods who ordain the concept of free will, and you succumb to its bribes of heaven, and its threats of hell (the equivalent of the old carrot-or-stick choice imposed on a mule to get it to do our bidding), then you probably are someone who was indoctrinated early in that belief, and, or, you are someone for whom the idea of spending a lazy eternity in heaven has a powerfully addictive appeal. Unfortunately, there can be no such thing as your being able to choose other than what you choose, no such thing as free will; nor can your god honestly claim to have given it to you, or to itself; the physical brain you inherited from your parents, and the well-meaning indoctrinations they imposed on your innocent, impressionable mind, and all the incidents that occurred in your life, and all the prior choices you've made with it, including those that occur the very moment before you make a new choice, dictate the choice; but, if you should be fortunate enough to have inherited a good brain; and if all those influences you have been subject to from the moment of your birth, will have been, in the main, good influences; and, if there is an actual heaven, and you have a belief that there is; you may very well find yourself there one fine day; though it won't be because you had a will that was free to choose good, rather then bad, but it will be because of your good fortune to have been born within a caring, embracing family and society that provided those necessary good influences.

Speaking of heaven, reminds me, of the tennis acquaintance I spoke of earlier, and his question of whether these issues of cause and effect, and determinism and free will, would apply in the eternal heaven he said he was sure to be occupying at some point. I told him that the concept of eternity, as it applies to the heaven he was anticipating, was beyond my comprehension. How long in time would a second be there? Would each Earth moment be stretched out eternally, or would there just be an eternity of individual moments? And so I had no idea about how eternity might impact the tiny truth I've described in this little treatise. But, for some reason I was unable to stop there, I rambled on about a subject I knew nothing about, and told him, that he should become acquainted with the term, light-years, because that would proba-bly be the only descriptive term they would use there in heaven; and to consider bringing sun-glasses and a flashlight with him since nobody seemed to know whether it would be day or night there. Those were the last words I spoke to him, I never saw him again. If he is dead now, I hope he was given what he was promised.

Finally, on that question of religion and its advocacy for free will: it is difficult to grasp why some religious founders would distort the truth, and assert that their god gave free will to

Adam and Eve, its' first human creations, when those founders must have been intelligent enough to have known that if their god was the all-knowing god they believed it to be, that god would know the impossibility of free will. What seems, more likely, is that their god did not make that impossible assertion; it may be more likely that it was the fault of some errant, ancient scribe who wrote down what he mistakenly thought he had heard his god say; or what he had wished his god to have said. Or perhaps that scribe took it upon himself, to make that claim of free will as a means of justifying what he thought was his god's, otherwise unjust, over-reaching punishment of the innocent, childlike, naive, trusting, and curious, Adam and Eve —along with the castigation of all their progeny— for committing a so-called sin, a sin because, they—of their own free will— disobeyed him and ate the fruit of a tree with the enticing name of *The Tree of Knowledge*. But, as I've said in a prior context, let's not waste our time wrestling with ourselves over the truth, or falsity, of the Adam and Eve fable; it is not worth the effort; we are all vested in what we believe of it; we have been taught it by those who had influence in our young lives, and by what we subsequently learned, on our own, from our experience of the social environment that we navigated in —all modulated by the hormones that are, at every moment, expressing themselves within us— and all that, together, will always attend the degree to which we believe that ancient fable of Adam and Eve and the supposed free will that their supposed God gave them; as it will, also, the degree of empathy we feel for our fellow man; and it will direct us in the choices, good or bad, that we make at every moment of our lives.

And, speaking of empathy, there are some of us who don't possess any empathy towards our fellow man regardless of our views on heaven and hell, either because we were born with a physically damaged brain incapable of empathy for others; or because we were indoctrinated —by word or by example— with the cynical idea that our fellow humans are not worthy of our concern; or because our minds were traumatically damaged by having experienced, or witnessed, horrific physical or verbal abuse that produced in us an unempathetic view of others —as might the possibility that we grew up in a very harsh, very competitive environment where our very survival depended on winning, at all costs, the daily battle with others. When we see people born deformed in mind or body, we generally accept that their inabilities are not their fault, that they are not responsible for their failings; we feel sorry for them, we feel empathy for them, we try to help them —we may even feel some empathy to those unfortunates with damaged brains, mentioned above, who lack any sense of empathy— but when we see people who grow up handsome and sturdy, and apparently healthy in body and mind, make failures of their lives, we denigrate them for that failure; we accuse them of being spoiled, of being physically lazy and mentally indolent, we say that they could have made a good life for themselves if only they had gotten up off their duffs and made the attempt, if only they had stopped hanging around without purpose, and done something useful with their lives. For those people, we have no pity, no empathy —especially if we, ourselves, once had some drawback we overcame— but they too are in need of empathy, for they were created with advantages but without the full complement of mental means or necessary conditions that would have allowed them to live the happy, productive life that was hoped for them. The truth is that we all come to adulthood crippled in one way or another; meaning that there are no perfect humans among us. Not that we can agree on what a perfect human would be like. Would it be female, or would it be male? Would it be straight, or of the LGBTQ community? Would it be pinkish, or would it be brownish, or reddish, or yellowish, or even brushed aluminumish? Would it be tall or would it be broad? Would it be an adventurous seeker of progress or a content seeker of sameness? Would the perfect human be the one who bows to the sacred or the one that stands up for reason? Would the perfect human be prudent, or would it be otherwise? Would the perfect human be ancient and experienced, or young and untested? Would the perfect human be someone revolted by the idea that they are a kissing cousin to the ape—one who may resent even being thought of as an animal? And, would this perfect human's index finger be the longest finger in its hand, or would its middle finger be the longest, or for that matter, the ring finger? Would it devour spinach and broccoli for nutrients, or would donuts satisfy its needs, or would it prefer to devour raw, blood-dripping meat as did its rugged, ancient ancestors? In actuality, many of us probably believe we know what the perfect human may be like, we may even, somehow, come to believe that the perfect human, or a close relative of it, is the one that stares at us in the mirror. But maybe not, maybe our egos are too passive to make that outrageous claim. So the best we can do for our imperfect, perfect selves is to strive mightily, as best we can, to tamp, or elevate, our egos, as needed, and do the very best we can to make ourselves and our world better, but to accept that none of us are the perfect human, that we all have flaws, and so, emulate a well known icon of yore, who, acknowledging his limitations and his attributes, was wont to proudly say,

"I yam what I yam, I'm Popeye, the sailor man,"
for we too are what we are, and the world
is what it is; and the best we can do is to
embrace that truth, so that we can know, when
we are dying—as we all must—that whether
we are lying in a prison cell, a rented room,
or a mansion in the clouds, we did
the best we could with what we
were given by the life and the
world we inherited.

No *ifs, ands*, or *buts*, about it.

www.ingramcontent.com/pod-product-compliance
Lightning Source LLC
Chambersburg PA
CBHW051734170526
45167CB00002B/925